THE WONDER OF DINOSAURS

THE WONDER OF DINOSAURS

Published by Fog City Press,
a division of Weldon Owen Inc.
415 Jackson Street
San Francisco, CA 94111 USA

www.weldonowen.com

weldon**owen**
President, CEO Terry Newell
VP, Publisher Roger Shaw
Associate Publisher Mariah Bear
Project Editor Bridget Fitzgerald
Creative Director Kelly Booth
Art Director Meghan Hildebrand
Text Roger Shaw
Production Director Chris Hemesath
Associate Production Director Michelle Duggan
Consultant Dr. Darren Naish

Weldon Owen is a division of **BONNIER**

Library of Congress Control Number on file with the publisher.

ISBN 13: 978-1-61628-791-7
ISBN 10: 1-61628-791-8

10 9 8 7 6 5 4 3 2

2015 2016

Printed in China.

A long, long time ago—millions of years before any humans existed—an amazing creature called the dinosaur ruled the earth. Dinosaurs were many different shapes, colors, and sizes.

All dinosaurs share one trait—they were all born from eggs, just like birds and snakes are today. Some baby dinosaurs grew up to be small, bird-like dinosaurs with feathers, and some became huge, ferocious dinosaurs with sharp teeth and claws.

There were about a thousand different types of dinosaurs. Most dinosaurs only ate plants—even the huge Apatosaurus.

Styracosaurus

Apatosaurus

Fun Fact

The word dinosaur means "terrible lizard!"

Stegosaurus

Ginkgo Tree
Fun Fact
The Jurassic and Cretaceous periods had gingko trees.

Pine Tree

Horsetails

Some of the plants that dinosaurs ate, such as pine trees and ginkgos, still exist today.

Fern Fossil

Fun Fact

Dinosaurs ruled the Earth for over 160 million years.

Fossil

Other types of plants and sea life from the age of the dinosaurs can still be seen in ancient fossils.

Tenontosaurus & Deinonychus

Some dinosaurs only ate meat and would use their curved claws and sharp teeth to kill other dinosaurs.

Allosaurus Claw

Deinonychus

Deinonychus

Deinonychus

Deinonychus had a talon on each hind foot called the "terrible claw."

Triceratops

Fun Fact

Over time, fossils become more like rock than bone.

Fossil Brush

Fossil Brush

Anything that once lived can leave behind traces. Scientists use special tools to remove fossils from the ground.

Tools with Mammoth Tusk

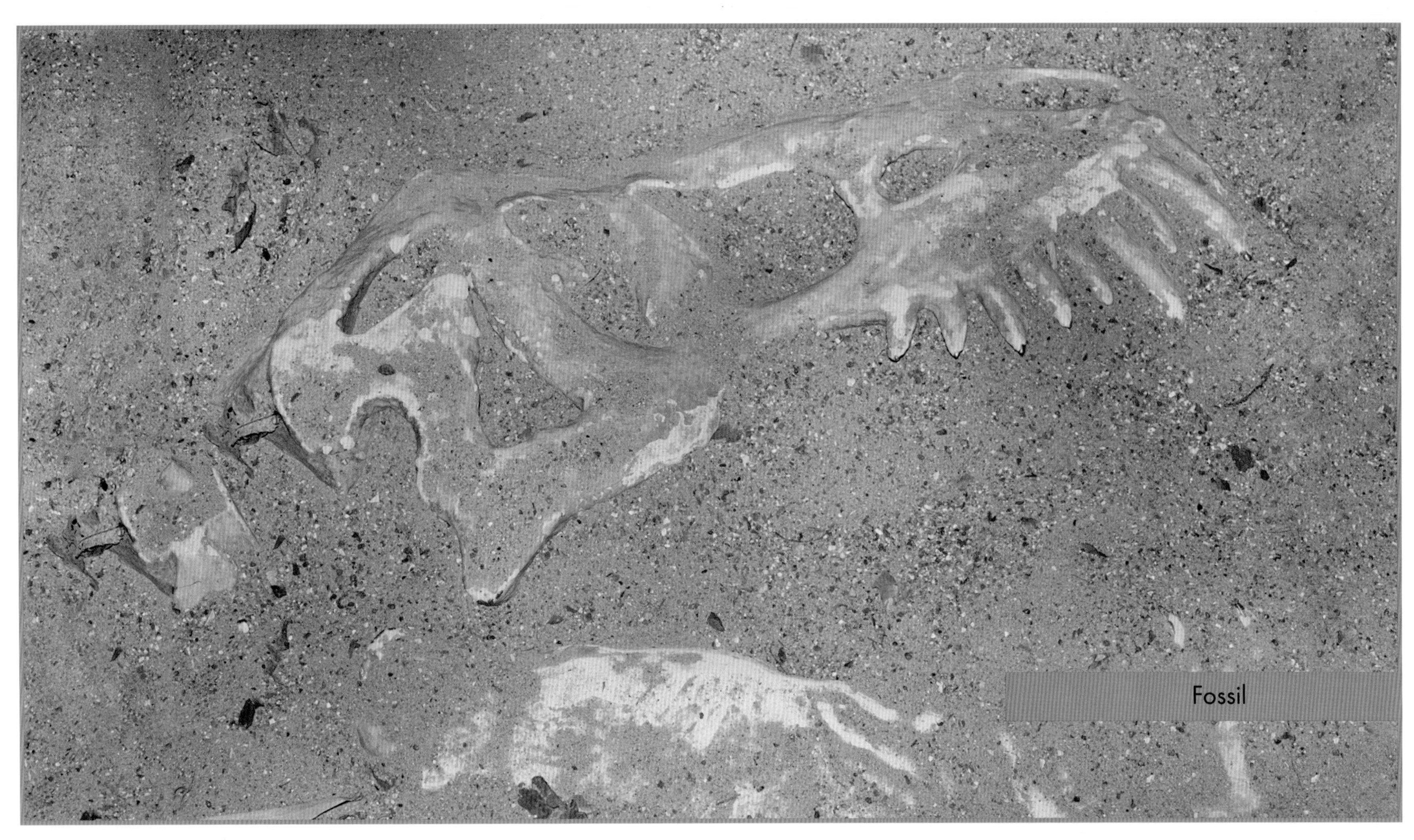
Fossil

Everything we know about dinosaurs is based on the study of fossils found all over the world.

Skull Fossil

Fun Fact

Most skulls were fragile, so they're very hard to find.

Fun Fact

Tyrannosaurus Rex had the biggest teeth!

Museum of Natural History

Tyrannosaurus Skeleton

Museum of Natural History

You can see dinosaur fossil skeletons in natural history museums. Imagine what it was like when huge dinosaurs roamed the earth!

Fun Fact

Hypsilophodons were very fast dinosaurs.

Hypsilophodon

Bambiraptor

Scientific research shows that many dinosaurs were very colorful and feathery, like birds.

Fun Fact

The name Oviraptor means "egg thief!"

Baby Oviraptor

Oviraptor

Hatching Oviraptor

The Oviraptor created nests and looked after its eggs in the same way that birds do today.

Fossilized eggs, some more than 100 million years old, tell us about how and where dinosaurs lived.

Fossilized Dinosaur Egg

Fossilized Eggs

Fun Fact

The smallest egg found is only 1 inch (3 cm) long.

University of Zurich Fossil

Kentrosaurus

Some dinosaurs had protective plates and spikes, like armor. Others had long claws or beaks.

Fun Fact
These long claws were used to pick fruit from trees.
Therizinosaurus

Brachiosaurus

Diplodocus

Carnotaurus

Some used their
huge size as
an advantage,
while others, like
Carnotaurus, had
horns on its head.

Fun Fact

Dinosaurs lived at the same time as many reptiles.

Pteranodon

Pteranodon

A close relative of dinosaurs, the Pteranodon was actually a flying, fish-eating reptile.

Dinosaur Footprint

All over the world, fossilized footprints of dinosaurs can be found embedded in rocks.

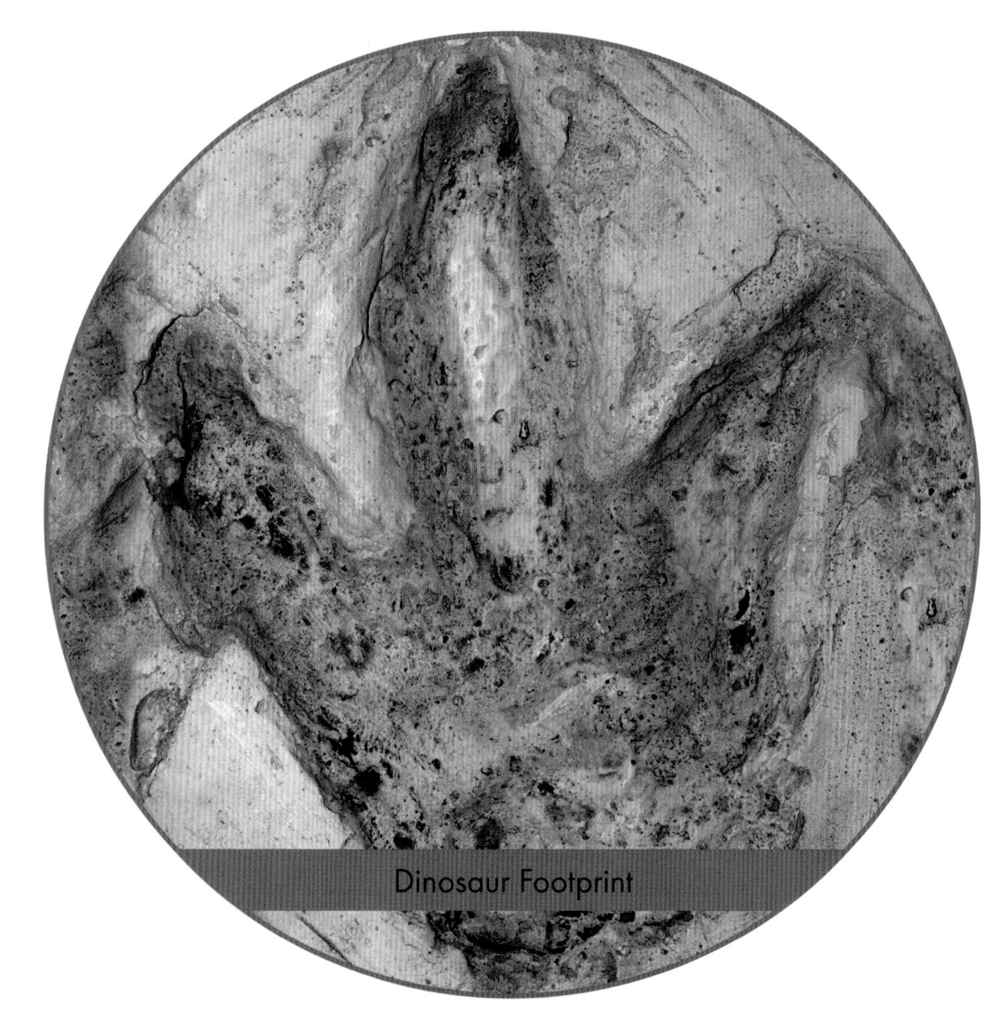
Dinosaur Footprint

Fun Fact

Chinese villagers used fossil bones in medicines!

Footprints in Namibia

Dilong Paradoxus

Fun Fact

This dinosaur lived 125 million years ago.

Sculptor Alan Groves

Artists build models for museums so that we can see what dinosaurs were like.

Fun Fact

The Stegosaurus had a brain the size of a walnut.

Diplodocus

Tyrannosaurus Rex

They also work with dinosaur scientists, or paleontologists, to create images of how dinosaurs may have looked.

Stegosaurus

Triceratops and Tyrannosaurus

You can also encounter amazing, life-sized dinosaurs in theme parks and museums around the world.

Triceratops

Fun Fact

The T. Rex was one of the biggest carnivores.

Uinta Mountains

You can visit places where many fascinating dinosaur fossils were discovered, too.

Dinosaur, Colorado

Fun Fact

Stegosaurus fossils were discovered in Colorado!

We still don't know why the dinosaurs disappeared 65 million years ago.

Comet

Mysterious Event

Volcanic Eruption
Fun Fact
Maybe there was
a disaster or fire.
It's a mystery!

Giganotosaurus

Key: t=top; b=bottom; AG=© Alan Groves, Art Dinouveau Pty Ltd; DM=© Damir G. Martin, Pixelmind; CR=© Chris Rallis, The Bright Agency; SS=Shutterstock.

Cover XX; 2, 5 DM; 7 AG; 8, 9 DM; 10, 11, 12, 13 SS, 14t AG; 14b DD; 15, 16 AG; 17, 18, 19, 20, 21 SS; 22 Jorg Hackemann/SS; 23 SS; 24, 25, 26, 27 AG; 28, 29, 30 SS; 31, 32, 33 DM; 34, 35 AG; 36, 37 SS; 38, 39 AG; 40, 41 CR; 42t SS; 42b Goran Cakmazovic/SS; 43 Elinag/SS; 44, 45, 46t SS; 46b DM; 47 SS; 48 DM.